# Insects
### and
# *Spiders*

# Moths and butterflies

## Shane F McEvey
### for the Australian Museum

This edition first published in 2002 in the United States of America by Chelsea House Publishers, a subsidiary of Haights Cross Communications.

Chelsea House Publishers
1974 Sproul Road, Suite 400
Broomall, PA 19008-0914

The Chelsea House world wide web address is www.chelseahouse.com

Library of Congress Cataloging-in-Publication Data Applied for.

ISBN 0-7910-6598-7

First published in 2001 by
Macmillan Education Australia Pty Ltd
627 Chapel Street, South Yarra, Australia, 3141

Edited by Anna Fern
Text design by Nina Sanadze
Cover design by Nina Sanadze
Australian Museum Publishing Unit: Jennifer Saunders and Catherine Lowe
Australian Museum Series Editor: Deborah White

Printed in China

**Acknowledgements**
Our thanks to Martyn Robinson, Max Moulds and Margaret Humphrey for helpful discussion and comments.

The author and the publisher are grateful to the following for permission to reproduce copyright material:

Cover: A Hercules moth, photo by Dominic Chaplin/Nature Focus.

Australian Museum/Nature Focus, pp. 6–7, 26, 27; Bill Belson/Lochman Transparencies, p. 30; C. Andrew Henley/Nature Focus, p. 29; Densey Clyne/Mantis Wildlife, pp. 5 (bottom), 9, 16, 18 (top and bottom), 19 (middle and bottom), 21 (top), 22 (top), 24 (top and bottom); Dominic Chaplin/Nature Focus, pp. 10 (top), 20 (bottom); Greg Harold, p. 7 (top); Gregory Wane/Nature Focus, p. 28; Hans & Judy Beste/Lochman Transparencies, pp. 14 (top and bottom), 17 (bottom); Jim Frazier/Mantis Wildlife, pp. 19 (top), 25 (bottom); Jiri Lochman/Lochman Transparencies, pp. 10 (bottom), 12 (top and bottom), 13 (top), 15 (bottom), 20 (top), 22 (bottom); John Cooper/Nature Focus, p. 15 (middle); John Kleczkowski/Lochman Transparencies, p. 13 (middle); Michael Cermak/Nature Focus, p. 23 (middle); Paul Zborowski, pp. 8 (top), 17 (top), 20 (middle), 23 (bottom); Pavel German/Nature Focus, pp. 8 (bottom), 11, 21 (bottom); Peter Marsack/Lochman Transparencies, pp. 5 (top), 13 (bottom), 15 (top), 25 (top); Steven David Miller/Nature Focus, p. 23 (top); T. & P. Gardener/Nature Focus, pp. 4, 18 (middle).

# Contents

## Glossary words

When a word is printed in **bold** you can look up its meaning in the Glossary on page 31.

# What are moths and butterflies?

Moths and butterflies are insects. Insects belong to a large group of animals called invertebrates. An invertebrate is an animal with no backbone. Instead of having bones, moths and butterflies have a hard skin around the outside of their bodies that protects their soft insides.

Moths and butterflies have:
- six legs
- four wings
- two **antennae**
- two eyes
- a mouth
- a long tongue
- many breathing holes on the sides of their bodies.

A female Cairns birdwing butterfly sits on a flower. The brown-colored female is bigger than the green-colored male.

# What makes moths and butterflies different from other insects?

Moths and butterflies have scales on their wings. These scales are like flattened hairs.

This is a close-up photo of the scales on the wings of a butterfly. These scales can come off on your fingers like dust if you press too hard. You can use a magnifying glass to see their flat shape.

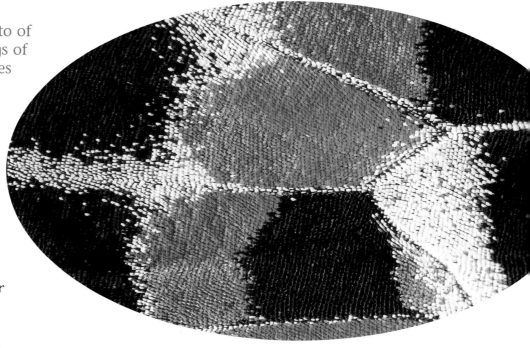

**Did you know?**

Butterflies are just another kind of moth. People notice butterflies because they are brightly colored and fly during the day.

Scientists have given a special name to all moths and butterflies. They are called **Lepidoptera**. Lepidoptera is a very large group and contains many different kinds of moths. There are three kinds of Lepidoptera that fly during the day. They are called day-flying moths, skippers and butterflies.

Not all moths are dull and not all butterflies are colorful. Many moths, like this tropical day-flying moth, are brightly colored. All butterflies have knobs on the ends of their antennae. One of the reasons that this day-flying moth is not a butterfly is that it does not have knobs on its antennae.

# Moth and butterfly bodies

The body of an adult moth or butterfly is divided into three segments. These segments are called the head, the **thorax** and the **abdomen**.

Adult moths and butterflies have scales all over their bodies, including around their mouths. These scales have many different shapes and sizes. The scales give the moth or butterfly its color.

## Head

On the head are the:
- mouth with its long tongue
- antennae
- eyes.

This is a regent skipper butterfly.

head

thorax

## Thorax

On the thorax are:
- six legs
- four wings
- breathing holes.

Some moths look very hairy, but these hairs are actually long scales.

abdomen

Fascinating Fact

In Australia, there are more than 11,000 kinds of moths and about 400 kinds of butterflies.

## Abdomen

The abdomen is where:
- food is digested
- females produce eggs
- males produce **sperm**
- breathing holes are.

# The head

On the head of an adult moth or butterfly are the mouth, eyes and antennae.

## Mouth

Moths and butterflies have a long tongue. They use their tongue to suck up liquid from flowers. They keep their tongue curled up when they are not drinking.

## Eyes

Moths and butterflies have **compound eyes**. This means that each eye is made up of lots of tiny eyes packed together.

## Antennae

Moths and butterflies use their antennae to feel and smell their environment. Their antennae can be simple, thread-like, feathery, or with a knob at the end.

tongue    eyes

This is a male Cairns birdwing butterfly.

## Did you know?

Moths and butterflies do not like flying when it is raining or windy. Instead, they like to sit quietly in a sheltered place like under a leaf or branch.

This Australian painted lady butterfly has simple, straight antennae with knobs on the ends.

# The thorax

On the thorax of an adult moth or butterfly are the legs, wings and some of the breathing holes.

## Legs

Moths and butterflies use their legs for landing and holding onto flowers while they feed.

## Wings

Moths and butterflies have two pairs of wings. There are two wings on each side of the body, even though sometimes it can look like there is only one wing on each side. Each wing contains a network of hard veins. These veins support the wings so they can be used for flying. Both pairs of wings are covered in scales.

## Breathing holes

Moths and butterflies breathe through tiny holes in the sides of their bodies called spiracles. Moths and butterflies do not breathe through their mouths.

### Fascinating Fact

Not all moths rest with their wings open and not all butterflies rest with their wings closed. Some moths, like the female case moth, do not even have wings. Many moths and butterflies constantly open and close their wings while they are feeding on flowers.

wings

legs

A dingy swallowtail butterfly sits on a flower. You can see the legs attached to the thorax and the abdomen extending out behind.

# Where do moths and butterflies live and what do they eat?

Moths and butterflies can live just about anywhere from the tropics to very cold places, from wet and humid rainforests to dry deserts, from mountains to the coast. Moths and butterflies live in certain places because that is where they find their food.

Hercules moth

## Adult moths and butterflies

Adult moths and butterflies drink the sweet liquid (called **nectar**) from flowers.

An adult moth feeds on wattle blossom. You can see that its tongue is uncurled and extended into the flower so that it can drink the nectar.

# Caterpillars

When moths and butterflies are young **caterpillars**, they eat many different plants. A caterpillar will usually spend all its time eating only one kind of plant. The kind of food the caterpillar eats depends on the kind of moth or butterfly it is. Some caterpillars eat wood. Others eat flowers, leaves, roots, fruit or seeds. Occasionally they will even eat other caterpillars if they are too crowded together.

## Fascinating Fact

The larvae of a clothes moth likes to eat animal hair, like wool. They can sometimes be found eating the woolen clothes in our closets.

An emperor gum moth caterpillar feeds only on the leaves of a eucalyptus tree. The caterpillar holds onto the plant with its six legs at the front end and its suckers at the back end.

# Moths and butterflies that live in deserts and dry habitats

Some moths and butterflies can live in very hot and dry places like deserts. Here are some of the moths and butterflies that can live in hot, dry places.

A checkered swallowtail butterfly drinking. Checkered swallowtails can live in dry, inland Australia.

Lesser wanderer butterflies live in dry areas. Their caterpillars like to eat a plant called milkweed. This butterfly is drinking nectar from a flower. Butterflies like to feed from flowers early in the morning before the day gets too hot.

These processionary moth caterpillars normally live all together in a web that they weave on their food plant. When they have eaten all the leaves on the plant, the caterpillars leave it to search for another plant to eat. They leave the plant and move along in one long line, each caterpillar closely following and even touching the one in front. When they find another plant they build another web to live in while they eat the plant.

Common Australian crow butterflies find shelter from the hot midday sun near the entrance of a cave. This is one way that butterflies that live in hot, dry areas avoid drying out.

## Did you know?

Moths and butterflies need to drink water when they live in hot, dry places.

Australian painted lady butterflies occur all over Australia, from the cities to the central deserts.

# Moths and butterflies that live in forests and wet habitats

Some moths and butterflies like to live in forests. Here are some of the moths and butterflies that live in forests and wet tropical places.

Fruit-sucking moths live in warm tropical forests and are a pest for fruit growers. The adult moth likes to suck the juice of ripening fruit. When they stick their tongue into the fruit, mold can get in. This causes the fruit to rot on the tree. Some of the fruit that fruit-sucking moths like to eat include pawpaws, lychees and citrus fruit such as oranges.

This hawk moth lives in tropical Queensland. As an adult it feeds on the nectar in rainforest flowers.

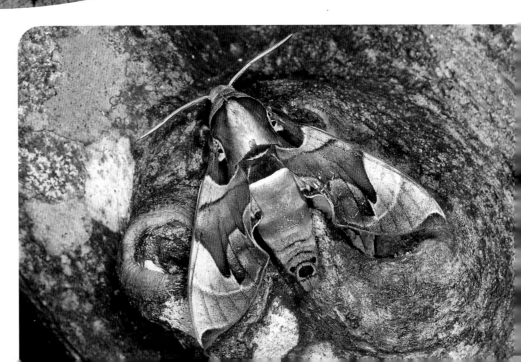

Yellow spot jewel butterflies live in mountainous areas. The caterpillars like to live on certain shrubs that grow in wet forests. These butterflies are active during the day.

Did you know?

Some plants, such as pawpaw, need hawk moths to visit them so that they can bear fruit. The hawk moth carries pollen from one flower to another. When the moth leaves pollen on a flower the flower will grow into a fruit.

Alpine skippers live in mountain valleys. The caterpillars feed on grasses that grow in mountain swamps. Adult alpine skippers are only active during the warm summer months.

Plume moths are unusual looking moths. They are not strong fliers and are usually found near their food plant. If you look closely, you can see short scales on this moth's wings and long scales on its legs. The plume moth often rests in this position. Its hind legs are sticking straight out behind it on either side of its abdomen. Plume moths are sometimes found around lights at night.

# How moths and butterflies communicate and explore their world

Moths and butterflies can get information about their environment in a number of ways. They can smell, feel and see.

Butterflies are active during the day and most use their eyes to find food and other butterflies. They can also use their antennae to feel their way around a flower and detect special smells from other butterflies.

Moths and butterflies are attracted to flowers by their color and perfume. This female wanderer butterfly is feeding on a sunflower.

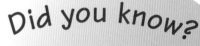

Most moths are active at night and find their food and other moths by smelling. Moths use their antennae to detect smells.

Moths and butterflies sometimes use their hair-like scales to feel surfaces and objects. Moths and butterflies also feel sounds. Moths can sense the sound of a bat. Bats eat moths.

# Moth and butterfly communication

Butterflies and moths can communicate in several ways.

- Being brightly colored helps day-flying moths and butterflies to find each other.
- Females release a special perfume from their bodies that attracts males.

Moths and butterflies do not communicate by making sounds.

Moths often fly at night. This hawk moth is flying around a flower of a rainforest tree.

Male moths and some butterflies find females by using their antennae to smell the perfume females produce. This male Hercules moth has large, feathery antennae so that he can detect this perfume.

17

# The life cycle of moths and butterflies

The whole life cycle of a moth or butterfly, from egg to adult, can take a few weeks or as long as a whole year.

Moths and butterflies reproduce **sexually**. This means that a male and a female are needed to make new moths or butterflies. The male provides sperm while the female provides eggs. The eggs and sperm need to join together for a new moth or butterfly to start growing. Males and females find each other by being attracted to the same places. This can be on a special plant, a bed of flowers or at the top of a hill. Sometimes the female attracts the male by giving off special smells.

The adult moth or butterfly eventually comes out of the pupa. After their wings expand, adult moths and butterflies do not grow any more. Most adult moths and butterflies only live for a few weeks.

An adult wanderer butterfly emerges from its pupal case. Its wings are still soft and small. The butterfly will stay sitting very still while it pumps its wing veins full of fluid. After the wings are fully outstretched, the veins harden. This whole process takes a couple of hours.

As the soft pupal skin gradually hardens the caterpillar becomes a **pupa**. In the **pupal stage**, they stay very still and do not eat — they are turning into adult moths or butterflies. Some caterpillars spin a silk **cocoon** around themselves before turning into a pupa.

The female lays many eggs that hatch into little caterpillars called **larvae**. These eggs are laid near or on food so that when the caterpillars hatch, they will have food to eat straight away.

The egg of a wanderer butterfly hatches and a tiny caterpillar crawls out. The caterpillar's first meal will probably be its egg case.

A male wanderer butterfly.

The little caterpillars spend the first part of their lives eating and growing bigger. As they grow, their skin becomes very tight until it splits. This allows the caterpillar to grow even bigger in its new, larger skin. This is called **molting** and it happens several times during the life of a caterpillar.

A wanderer butterfly caterpillar feeds on a milkweed leaf.

The pupa of a wanderer butterfly.

This wanderer butterfly caterpillar is about to shed its last larval skin. It has hooked its tail end to a pad of silk that was attached to the plant earlier. When it is a pupa it will hang from this pad of silk.

The last molt of the caterpillar is special. This is when the last caterpillar skin splits open, revealing a soft pupal skin underneath.

# Caterpillars

Caterpillars are very different from adult moths or butterflies because:

- they are worm-like and have big strong jaws
- they do not have wings or a three-segmented body
- they eat different food
- they do not travel large distances
- they are found in different places
- they are active at different times of the year
- they protect themselves in different ways.

Birdwing butterfly caterpillars eat vine leaves that grow in rainforests. These caterpillars can grow to over 10 centimeters (4 inches) long.

Caterpillars, like this looper moth caterpillar, look very different from the adult moth they will eventually become.

Caterpillars are usually found on their food. Sometimes a number of caterpillars will live together in a web of silk that they all weave together. When they make their web, the caterpillars weave irritating hairs from their bodies into it. This helps to protect the caterpillars from predators.

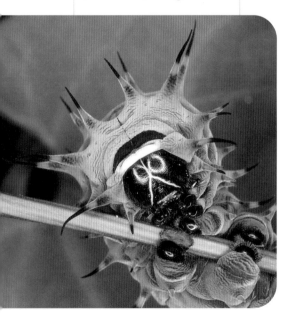

Case moth caterpillars build cases made of silk and sticks to live in. They carry the case with them wherever they go. When something attacks them, they hide inside.

The case moth caterpillar carries around a case of silk and sticks.

# Cocoons

Some caterpillars build a cocoon before they become a pupa. The cocoon protects the pupa while it is turning into an adult. Cocoons can be different shapes and sizes but are always made of silk. Sometimes caterpillars will build leaves and hairs into their cocoons for extra protection.

When this white-stemmed gum moth caterpillar made its cocoon, it wove hairs from its body into the silk. These hairs hurt to touch. This helps protect the pupa inside the cocoon.

# Pupae

The pupa of a moth or butterfly is usually hard. It has no legs and keeps very still. When the adult is ready to leave the pupa, it splits the pupal skin open.

When the adult moth or butterfly first comes out of the pupa, its wings are very small. If it is inside a cocoon, it makes a hole to escape. The adult will then sit still while it pumps liquid through the veins in its wings to make them expand. Once this is complete, the adult will fly off and leave the pupal skin behind.

## Did you know?

Witchetty grubs are moth caterpillars. One kind lives in the ground and eats roots. Sometimes you can find their large brown empty pupal skins poking out of the ground under acacias.

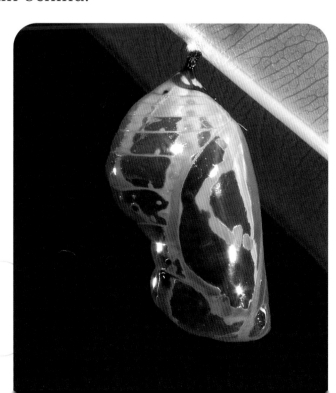

The pupa of the common Australian crow butterfly looks like gold.

21

# Predators and defenses

When moths and butterflies are caterpillars, they can be eaten by insect predators and **parasites**. Sometimes caterpillars are eaten by larger animals like birds, lizards and little mammals.

Moth caterpillars that live in the ground have to protect themselves from parasites such as **nematode worms** and **mites**. Adult moths and butterflies can also die from diseases and some types of **fungus**.

Adult moths and butterflies are eaten by other insects, bats, spiders, fish and birds.

This caterpillar is full of insect parasites. If you look closely you will see that the dead caterpillar's body is now packed full of little insect larvae.

This skipper has been caught by a jumping spider. The spider's poison paralyzes the muscles of the larger butterfly so it cannot struggle or fly away.

# Caterpillar defenses

Caterpillars can protect and defend themselves by:
- having spines and irritating hairs
- spitting fluid
- dropping towards the ground
- wriggling
- hiding
- being active only at night
- being hard to see.

Some caterpillars avoid being eaten by looking like something they are not. This giant swallowtail caterpillar has patterns on its body that look like snake eyes. This makes the caterpillar look dangerous.

The spines on this cup moth caterpillar sting. This helps protect the caterpillar from being eaten.

# Adult defenses

Adult moths and butterflies defend themselves by:
- flying away
- only being active at night (only moths are active at night)
- being hard to see.

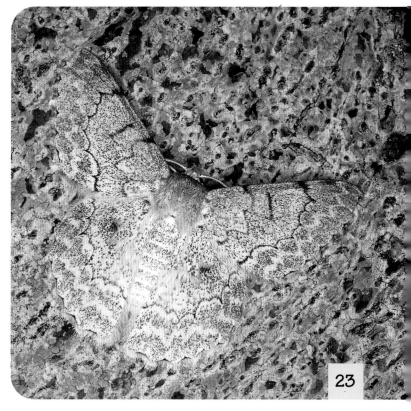

One of the ways that moths and butterflies avoid being eaten is by being hard to see. Can you see the moth in this photo?

# Weird and wonderful moths and butterflies

Welcome to the wonderful world of bizarre and extraordinary moths and butterflies!

## Caterpillar surprises

The caterpillars of swallowtail butterflies have special red horns that pop out from behind their heads. When the horns pop out they release a strong smell. This smell repels predators.

Swallowtail caterpillar

## Whistling moths

When the whistling moth flies, its wings make a whistling sound.

## Case moths

The adult female case moth has no wings, no legs, no eyes and practically no mouth. She lives deep inside a 'case' or house built of silk. The adult male finds her by a special perfume she releases. When the male finds her he has to extend his abdomen so that he can mate with her inside her silk house.

### Did you know?

Silk is made from cocoons of the silk moth. People have been farming silk moth caterpillars (often called silkworms) for thousands of years. The silk thread is collected by unwinding the cocoon built by the caterpillar. The thread is then woven into silk fabric.

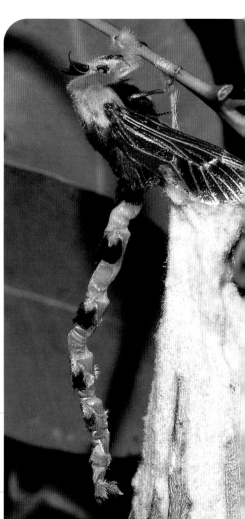

A male case moth with his abdomen extended.

## Migrating moths

In spring, Bogong moths fly south through Sydney and Canberra on their way to the Southern Alps, where they spend the summer. During summer, millions and millions of individual moths will pack together in rock crevices. Aborigines discovered that they were a delicious source of food and many would go up into the mountains to feast on the moths.

## Ant friends

These azure butterfly caterpillars are being cared for by ants. The ants get a sweet fluid from the caterpillars without harming them. In return, the ants protect the caterpillars from parasites.

Several kinds of butterfly caterpillars live inside ant nests. One type lives there because it eats ants. The caterpillar has a leathery skin that helps protect it from ant bites. The ants feed on a sweet liquid that they get from the side of the body of the caterpillar.

## Moth froth

This moth produces a frothy orange fluid out of its thorax. When this fluid comes out it makes a sizzling sound. This helps protect the moth from predators.

# Collecting and identifying moths and butterflies

There are so many different moths and butterflies that scientists are still discovering many new kinds, especially moths. If a scientist catches a moth or butterfly that is unknown, they name it and describe it so other scientists can study it too.

An example of each kind of moth and butterfly that has been found is kept in collections at museums. Some museums have thousands of different kinds of moths and butterflies in their collections and millions of different kinds of insects. These collections are used by scientists who want to study and learn about moths and butterflies.

When scientists collect moths and butterflies, they use special equipment. They sometimes use a bright light to attract the moths, nets to catch them and containers to put them in. Caterpillars are sometimes kept alive on their special food plant to see what they will become.

Moths and butterflies in museum collections are pinned or sometimes kept in paper envelopes. Each moth or butterfly should have a label with information about where and when it was collected, and who collected it.

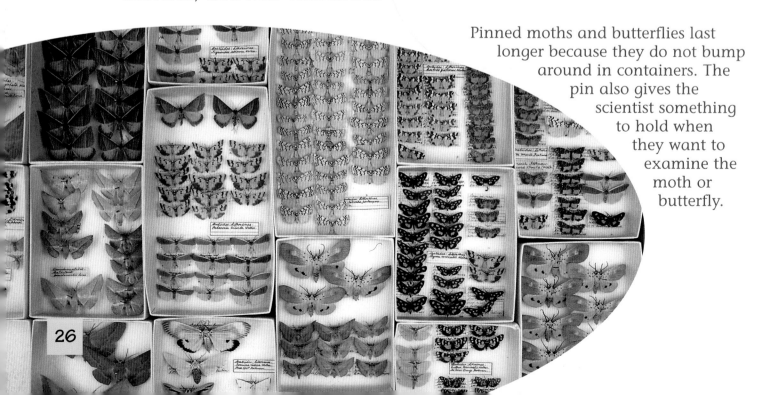

Pinned moths and butterflies last longer because they do not bump around in containers. The pin also gives the scientist something to hold when they want to examine the moth or butterfly.

# How are moths and butterflies identified?

Scientists identify moths and butterflies by looking very carefully at their shape, size and color. If a moth or butterfly's shape, size and color are different to all other moths and butterflies that are already known, then it is considered a new kind and is given a new scientific name.

# What do scientists study about moths and butterflies?

After a moth or butterfly has been given a name, scientists then study:

- where it lives
- what it eats
- how often it molts
- what makes it turn into a pupa or an adult
- what are its natural predators
- what poisons or pollutants kill it or interfere with its normal life cycle.

## Fascinating Fact

When studying moths and butterflies scientists pay special attention to the arrangement of veins in the wings. The veins help scientists identify what kind of moth or butterfly they are looking at.

A scientist studying butterflies and moths.

# Ways to see moths and butterflies

How many different kinds of moths and butterflies can you observe around you?

- Leave a light on outside at night in summer and look at the different kinds of moths that come to the light.
- In spring and summer watch butterflies visiting flowers.

The caterpillars of this beautiful male emperor gum moth can be found eating gum leaves.

# Looking for caterpillars

Look for bite marks on young leaves and see if you can find the animal that made them. It may be a caterpillar. You may have to look very carefully, as the caterpillar can be the same color as the leaf.

- Look on cabbage leaves for cabbage white caterpillars.
- Look on citrus trees for orchard swallowtail caterpillars.
- Look on gum leaves for emperor gum moth and many other moth caterpillars.
- Look on grapevine leaves for vine moth caterpillars.

**Did you know?**

No adult moth or butterfly can bite, sting or hurt people in any way but hairy caterpillars should be handled with care.

Cabbage white caterpillars eating a cabbage leaf.

# Moths and butterflies quiz

1 How many wings do moths and butterflies have?

2 What do moths and butterflies have on their wings?

3 What is the special name scientists have given all moths and butterflies?

4 Can adult moths and butterflies bite?

5 What do moths and butterflies do with their long tongue when they are not feeding?

6 Are moths and butterflies animals?

7 Do moths and butterflies have veins in their wings?

8 What do adult moths and butterflies eat?

9 Where does the largest moth in the world live and what is it called?

10 What do flowers use to attract moths and butterflies to them?

11 Why don't adult moths and butterflies eat leaves?

12 Why do some caterpillars build cocoons?

13 What are witchetty grubs?

14 What part of the whistling moth makes a whistling sound?

15 How do cup moth caterpillars protect themselves from predators?

**Check your answers on page 32.**

When strongly scented flowers start to bloom, look closely at them on a warm sunny evening around dusk. See if you can spot moths like this one feeding.

# Glossary

| | |
|---|---|
| **abdomen** | The rear section of the body of an animal. |
| **antennae** | The two 'feelers' on an insect's head that are used to feel and smell. (Antennae = more than one antenna.) |
| **caterpillar** | The larva of a moth or butterfly. |
| **cocoon** | A silk house made by a caterpillar to shelter in when it becomes a pupa. |
| **compound eyes** | Eyes that are made up of many tiny eyes packed together. |
| **defenses** | The ways that animals and plants protect themselves from predators. |
| **fungus** | Mushrooms and toadstools are kinds of fungus. Some fungi grow flat on or under the skin of animals. (Fungi = more than one fungus.) |
| **larvae** | Caterpillars, grubs and maggots are kinds of larvae. In the life cycle of an insect the larval stage is after the egg stage and before the pupal stage. Larvae hatch out of eggs, grow and then turn into pupae. (Larvae = more than one larva.) |
| **Lepidoptera** | The scientific name for moths and butterflies. |
| **mites** | Small spider-like animals that are not insects. |
| **molting** | When an animal sheds its entire skin it molts. The process is called molting. |
| **nectar** | A sweet fluid found in flowers. |
| **nematode worm** | A kind of worm (not an earthworm) that is usually pointed at the ends, shiny, wriggles in loops and is often smaller than 1 centimeter (0.39 inch) long. |
| **parasite** | An animal or plant that lives on or in another animal or plant. |
| **pupae** | In the life cycle of insects (like moths, beetles and flies) larvae turn into pupae. Adult insects later emerge from pupae. A cocoon is a shell around a pupa. (Pupae = more than one pupa.) |
| **pupal stage** | A stage in the life cycle of an insect when the insect is a pupa. |
| **sexual reproduction** | When a male and female living thing combine to make more living things. |
| **sperm** | The male reproductive cell. |
| **thorax** | The middle section of an animal's body. |

# Index

## Answers to quiz

1 four 2 scales 3 Lepidoptera 4 no 5 they curl it up 6 yes 7 yes 8 nectar 9 Australia and it is called a Hercules moth 10 color and smell 11 because they cannot bite — only caterpillars have mouths for biting and eating leaves 12 to protect them while they are a defenseless pupa turning into an adult 13 moth caterpillars 14 its wings 15 by having spines on their bodies.